Beyond the Horizon: Rethinking Progress

Dimensions: Time, Future

by Her Majesty Queen Dillys I

of the Royal Wright Family

Table of Contents

Introduction

We live in an age dominated by immediacy. Political cycles span mere years, corporate strategies pivot quarterly, and digital platforms refresh by the second. The present moment — urgent, noisy, and fleeting — has become the default horizon for decision-making. But what happens when we design only for now?

The consequences are visible everywhere: infrastructure that crumbles within decades, ecological systems pushed to collapse, and policies that fail to anticipate the needs of future generations. The tyranny of the present has narrowed our imagination, eroded our foresight, and compromised our legacy.

This book is a call to rethink progress through the lens of **time**. It explores how long-term thinking — sometimes called "deep time" planning — can reshape how we govern, build, and care for the planet. It draws on five regional case studies that exemplify this shift:

- The **Future Generations Act** in Wales, UK
- **Finland's Forest Strategy** in Europe
- **Vision 2071** in the United Arab Emirates
- **Ethiopia's Reforestation Movement** in Africa
- **GIFT City** in India, Asia

Each case offers a distinct approach to designing for futures that extend beyond the lifespan of a single administration, project, or generation. Together, they reveal a growing global movement: one that seeks to embed responsibility, resilience, and reverence into the architecture of progress.

This book is not a blueprint. It is a provocation — an invitation to stretch our civic imagination, to design with time as our ally, and to build a world worthy of inheritance.

Chapter 1: Time as a Design Principle

Time is not just a measurement — it is a medium. Like space, material, and form, time shapes how we design, govern, and relate to the world. Yet in modern systems, time is often treated as a constraint: a deadline, a budget cycle, a product launch window. Rarely is it embraced as a principle of design.

To rethink progress, we must first rethink time.

Philosophical Foundations

Across cultures, time has been understood in radically different ways. Western industrial societies tend to view time as linear — a forward march from past to future. Indigenous traditions often see time as cyclical, regenerative, and relational. In Māori cosmology, for example, ancestors are not behind us but in front — guiding our steps. In Buddhist thought, time is fluid, impermanent, and interconnected.

Designing for time means choosing which philosophy we honour. Do we build for the next fiscal quarter, or the next century? Do we prioritize speed, or stewardship?

Temporal Bias in Infrastructure

Most infrastructure is built with short-term horizons. Roads, bridges, and buildings are designed to last decades — not centuries. Yet some of the world's most enduring structures, from Roman aqueducts to Japanese temples, were built with deep time in mind. They reflect a commitment to legacy, not just utility.

Modern examples are rare but instructive. The Svalbard Global Seed Vault in Norway is designed to last 1,000 years, safeguarding biodiversity against planetary catastrophe. The

Long Now Foundation in the United States is building a 10,000-year clock — a monument to long-term thinking.

These projects challenge the dominant paradigm. They ask: *What would we build if we knew it had to last for generations?*

Designing for 50, 100, 500-Year Horizons

Long-term design requires new tools and mindsets. It demands:

- **Scenario planning**: imagining multiple futures, not just one forecast
- **Back casting**: starting with a desired future and working backward
- **Legacy metrics**: measuring impact across decades, not months

It also requires humility. We must accept that we cannot predict the future — but we can prepare for it. We can build systems that are adaptable, resilient, and rooted in values that endure.

Time is not a constraint. It is a canvas.

Chapter 2: The Future Generations Act (UK)

In 2015, the Welsh Government passed a groundbreaking piece of legislation: the **Well-being of Future Generations (Wales) Act**. It was the first law of its kind in the world — a legal commitment to consider the needs of future generations in every decision made by public bodies. In a political landscape often dominated by short-termism, this act was a radical gesture of long-term responsibility.

A Law for Tomorrow

The Future Generations Act requires public institutions — including local councils, health boards, and national agencies — to work toward seven well-being goals:

1. A prosperous Wales
2. A resilient Wales
3. A healthier Wales
4. A more equal Wales
5. A Wales of cohesive communities
6. A Wales of vibrant culture and thriving Welsh language
7. A globally responsible Wales

These goals are not abstract ideals. They are legally binding targets that must inform policy, budgeting, and service delivery. Every decision — from building a school to approving a transport plan — must be assessed for its long-term impact.

The Future Generations Commissioner

To ensure accountability, the Act created the role of the **Future Generations Commissioner for Wales**. This independent office serves as a watchdog, advisor, and advocate for future citizens.

The Commissioner reviews policies, challenges short-sighted decisions, and promotes best practices in long-term thinking.

Under the leadership of Sophie Howe (2016–2022), the office gained international recognition. Howe pushed for sustainable transport, climate action, and inclusive urban planning — often challenging entrenched interests and political inertia. Her successor continues this legacy, reinforcing the idea that future generations deserve a voice today.

Impact on Policy and Infrastructure

The Act has already reshaped how Wales approaches infrastructure and development. For example:

- **Transport**: The cancellation of the M4 relief road near Newport was influenced by the Act's principles. Instead of expanding car infrastructure, the government invested in public transport and active travel.
- **Housing**: New housing developments are now evaluated for long-term sustainability, energy efficiency, and community cohesion.
- **Education**: Schools are encouraged to teach future literacy — helping students understand climate change, civic responsibility, and ethical innovation.

These shifts are not always easy. Long-term thinking often clashes with short-term political cycles, budget pressures, and public impatience. But the Act provides a framework — a moral compass — that helps navigate these tensions.

Global Influence

The Future Generations Act has inspired interest from governments around the world. Delegations from New Zealand, Canada, and the European Union have studied its model. The

United Nations has cited it as a best practice in sustainable development governance.

But replication is not simple. The Act works because it is embedded in Welsh culture — a culture that values community, heritage, and stewardship. Exporting the model requires adaptation, not duplication.

Challenges and Critiques

Despite its promise, the Act faces challenges:

- **Enforcement**: While the goals are legally binding, enforcement mechanisms are limited. Some critics argue that the Act lacks teeth.
- **Political Will**: Long-term thinking requires courage — especially when it conflicts with short-term popularity.
- **Public Engagement**: The concept of future generations can feel abstract. Translating it into everyday language and action remains a challenge.

Yet these critiques do not diminish the Act's significance. They highlight the complexity of designing for time — and the need for continuous reflection and refinement.

Lessons for the World

The Future Generations Act teaches us that long-term thinking is not just a philosophical ideal — it can be legislated, institutionalized, and practiced. It shows that governments can be stewards, not just managers; architects of legacy, not just operators of systems.

It also reminds us that the future is not a distant abstraction. It is shaped by every decision we make today — every road we build, every law we pass, every value we uphold.

In Wales, the future has a seat at the table. The question is: *Can the rest of the world follow suit?*

Chapter 3: Finland's Forest Strategy (Europe)

In the heart of Northern Europe, Finland stands as a model of ecological foresight. With over 75% of its land covered in forest, the country has long understood that trees are more than timber — they are infrastructure, identity, and inheritance. Finland's forest strategy is not just about managing resources; it's about designing a future that is biodiverse, resilient, and deeply rooted in place.

Forests as Ecological Infrastructure

Finland's forests are living systems that provide climate regulation, water purification, carbon sequestration, and cultural continuity. They are home to thousands of species and serve as carbon sinks in a region increasingly affected by warming temperatures.

The Finnish government treats forests as **ecological infrastructure** — assets that must be maintained, restored, and protected. This approach reframes forestry from extraction to stewardship. It recognizes that forests are not just economic commodities but planetary partners.

The Long-Term Strategy

Finland's **National Forest Strategy 2035** outlines a vision for sustainable forest use that balances economic, ecological, and social goals. Key pillars include:

- **Biodiversity protection**: Expanding conservation areas and restoring degraded habitats
- **Climate mitigation**: Enhancing carbon storage through afforestation and reduced clear-cutting

- **Sustainable timber production**: Promoting selective logging and renewable materials
- **Citizen engagement**: Encouraging public participation in forest planning

The strategy is informed by scientific modeling, stakeholder consultation, and intergenerational ethics. It asks not only what forests can do for us today, but what they must be for future generations.

Citizen Participation and Cultural Identity

Forests are deeply woven into Finnish identity. They appear in folklore, literature, and national rituals. Many Finns own small forest plots, passed down through families. This personal connection fosters a culture of care and responsibility.

Public participation is central to forest governance. Citizens are invited to contribute to planning processes, attend forest walks, and engage in biodiversity monitoring. This democratization of stewardship ensures that forests are not managed solely by technocrats — but by communities.

Lessons for the World

Finland's forest strategy offers several lessons for global ecological planning:

- **Design with time**: Forests grow slowly. Planning must span decades, not years.
- **Balance interests**: Economic use and ecological protection can coexist — but only with clear boundaries and shared values.
- **Engage the public**: Stewardship is strongest when it is participatory.

- **Honor culture**: Conservation is more effective when it resonates with identity.

In an age of deforestation and climate crisis, Finland reminds us that forests are not relics of the past — they are blueprints for the future.

Chapter 4: Vision 2071 (UAE)

In the United Arab Emirates, the future is not a distant abstraction — it is a national priority. The UAE's **Vision 2071** is a long-term strategic framework that reimagines the country's trajectory over the next half-century. Launched in 2018 to mark the centennial of the nation's founding, Vision 2071 is a bold declaration: that the UAE intends not only to adapt to the future, but to shape it.

Strategic Foresight as Governance

Vision 2071 is rooted in the principles of **foresight governance** — the idea that governments must anticipate future challenges and opportunities, not merely react to them. It builds on the earlier **UAE Centennial 2071** initiative, which emphasized investment in education, innovation, and national identity.

The strategy spans four key pillars:

1. **Education**: Preparing future generations with skills in AI, robotics, and ethics
2. **Economy**: Transitioning to a knowledge-based, diversified economy
3. **Government**: Building agile, data-driven institutions
4. **Society**: Preserving culture while embracing global citizenship

This vision is not confined to policy documents. It informs urban planning, infrastructure development, and diplomatic strategy. It is embedded in the architecture of the nation.

Designing for the Far Future

Vision 2071 challenges conventional planning horizons. Most national strategies span 5 to 10 years. Few dare to look 50 years ahead. But the UAE's approach is deliberate: by extending the timeline, it creates space for **transformational thinking**.

For example:

- **Space exploration**: The UAE has launched the Hope Probe to Mars and plans to build a human settlement on the Red Planet by 2117.
- **AI governance**: The country appointed the world's first Minister of Artificial Intelligence, signaling a commitment to ethical tech leadership.
- **Smart cities**: Projects like Masdar City aim to model sustainable urban living in desert environments.

These initiatives reflect a belief that the future must be **designed**, not deferred.

Cultural Preservation and Innovation

Long-term planning in the UAE is not only about technology — it is also about **identity**. Vision 2071 emphasizes the importance of preserving Emirati culture, language, and values. This includes:

- Reviving traditional crafts and architecture
- Promoting Arabic digital content
- Supporting intergenerational storytelling and heritage education

In a rapidly globalizing world, the UAE seeks to balance **modernity with memory**. It recognizes that progress must be rooted in place — and that cultural continuity is a form of resilience.

Leadership and Legacy

Vision 2071 is driven by leadership that understands the power of narrative. By articulating a clear and ambitious future, the UAE inspires public engagement, institutional alignment, and international collaboration.

But the vision also raises questions:

- Can long-term strategies remain flexible in a volatile world?
- How can citizens participate meaningfully in shaping distant futures?
- What mechanisms ensure accountability across generations?

These questions are not unique to the UAE. They apply to any nation seeking to govern with foresight.

Lessons for the World

Vision 2071 offers a compelling model of **future-oriented governance**. It shows that:

- Long-term thinking can be institutionalized
- Technology and culture can co-evolve
- Strategic foresight can inspire national purpose

In a world often paralyzed by short-term crises, the UAE reminds us that the horizon is not a limit — it is an invitation.

Chapter 5: Ethiopia's Reforestation Movement (Africa)

In the highlands and savannas of Ethiopia, a quiet revolution is taking root — quite literally. The country's ambitious reforestation campaign, known as the **Green Legacy Initiative**, is one of the largest ecological restoration efforts in the world. It is not merely a tree-planting program; it is a generational act of renewal, resilience, and responsibility.

The Green Legacy Initiative

Launched in 2019 under the leadership of Prime Minister Abiy Ahmed, the Green Legacy Initiative set out to plant **20 billion trees** by 2024. The goal was not only to combat deforestation and desertification, but to restore degraded landscapes, improve food security, and mitigate climate change.

In its first year alone, Ethiopia planted over **4 billion trees**, mobilizing millions of citizens across urban and rural areas. Schools, churches, mosques, and community groups participated in coordinated planting days. The campaign became a national ritual — a symbol of unity and hope.

Reforestation as Climate Strategy

Ethiopia faces severe environmental challenges: soil erosion, erratic rainfall, and declining agricultural productivity. Reforestation is a strategic response to these threats. Trees stabilize soil, regulate water cycles, and sequester carbon — making them essential allies in climate adaptation.

But the Green Legacy Initiative goes beyond ecology. It is embedded in Ethiopia's **national identity**. It draws on indigenous knowledge, local species, and community stewardship. It treats forests not as commodities, but as commons.

Community-Led Planning

One of the initiative's strengths is its **bottom-up approach**. While the government provides coordination and resources, local communities lead the planning and implementation. Farmers choose species based on soil type, water availability, and cultural significance. Elders guide planting rituals. Youth groups monitor growth and survival rates.

This participatory model ensures that reforestation is not imposed — it is embraced. It builds ownership, pride, and continuity.

Ecological and Economic Impact

The benefits of reforestation are already visible:

- **Improved soil fertility** and crop yields
- **Reduced flooding and drought vulnerability**
- **New sources of income** from fruit trees and agroforestry
- **Revived biodiversity** in degraded habitats

These outcomes demonstrate that ecological restoration can be economically viable — especially when designed with long-term horizons.

Challenges and Critiques

Despite its success, the initiative faces challenges:

- **Survival rates**: Not all planted trees thrive. Maintenance and monitoring are critical.

- **Land tenure**: Unclear property rights can hinder long-term stewardship.
- **Species selection**: Some critics warn against monoculture or non-native species.
- **Political continuity**: Will future administrations sustain the momentum?

These challenges are not unique to Ethiopia. They reflect the complexity of designing for deep time — where success depends not just on planting, but on **cultivating**.

Lessons for the World

Ethiopia's Green Legacy Initiative offers powerful lessons:

- **Scale is possible** when rooted in culture and community
- **Ecological planning** must be participatory and place-based
- **Reforestation** is not just environmental — it is civic, economic, and spiritual
- In a world grappling with climate collapse, Ethiopia reminds us that restoration is not a dream — it is a duty. And it begins with a single seed.

In a world grappling with climate collapse, Ethiopia reminds us that restoration is not a dream — it is a duty. And it begins with a single seed.

Chapter 6: GIFT City and India's Urban Futures (Asia)

In the western state of Gujarat, India is building a city designed for the future — a financial and technological hub known as **GIFT City** (Gujarat International Finance Tec-City). Conceived as a smart, sustainable, and globally integrated urban center, GIFT City is more than a real estate project. It is a statement of intent: that India is ready to lead in shaping the urban futures of the 21st century.

A City Designed for Tomorrow

GIFT City was envisioned to address a critical gap in India's infrastructure: the absence of a world-class financial district that could rival global centres like London, New York, or Singapore. But its ambition goes beyond finance. The city is designed to be:

- **Smart**: Equipped with integrated ICT systems for traffic, utilities, and governance
- **Sustainable**: Featuring district cooling, waste-to-energy systems, and green buildings
- **Inclusive**: Offering housing, education, and healthcare within walking distance of workplaces
- **Globally connected**: Operating under a special economic zone with international regulatory standards

This convergence of technology, policy, and planning positions GIFT City as a prototype for future Indian cities — cities that are efficient, resilient, and globally competitive.

Infrastructure as a Time Machine

Urban infrastructure is often reactive — built to solve yesterday's problems. GIFT City flips this paradigm. It is designed with

anticipatory logic, incorporating features that prepare for future challenges:

- **Climate adaptation**: Flood-resistant drainage, heat-resilient materials
- **Mobility innovation**: Electric vehicle infrastructure, walkable districts
- **Digital governance**: E-services, biometric access, and real-time data dashboards

These systems are not just technical upgrades — they are expressions of **temporal design**. They reflect a commitment to building cities that endure, evolve, and empower.

Economic Inclusion and Global Integration

GIFT City aims to democratize access to global finance. By offering regulatory clarity, tax incentives, and digital platforms, it seeks to attract startups, investors, and institutions from across India and abroad.

But inclusion is not automatic. Critics warn that GIFT City may replicate patterns of exclusion seen in other elite urban enclaves — where access is limited to the affluent, and surrounding communities are marginalized.

To fulfill its promise, GIFT City must ensure:

- Affordable housing and public services
- Transparent governance and civic participation
- Integration with regional economies and cultures

Urban futures must be **inclusive**, not insular.

Ecological Footprint and Urban Ethics

Smart cities often promise sustainability — but their ecological footprints can be significant. GIFT City's construction has raised concerns about land use, biodiversity loss, and water consumption.

Designing for the future requires more than green branding. It demands:

- **Ecological accounting**: Measuring impact across decades
- **Regenerative planning**: Restoring ecosystems, not just minimizing harm
- **Cultural sensitivity**: Honouring local landscapes, traditions, and communities

Urban ethics must extend beyond efficiency. They must embrace **care**.

Lessons for the World

GIFT City offers a glimpse into the future of urban India — and into the global challenge of designing cities that are smart, sustainable, and just. It shows that:

- Long-term urban planning is possible
- Technology can enhance governance and resilience
- Economic ambition must be matched by ecological and social responsibility

In a century of rapid urbanization, GIFT City reminds us that the cities we build today will shape the lives of billions tomorrow. The question is not whether we can build smart cities — but whether we can build **wise ones**.

Chapter 7: Designing for Deep Time

Most design operates within narrow timeframes — product cycles, election terms, quarterly reports. But some challenges demand a longer view. Climate change, biodiversity loss, cultural erosion, and intergenerational justice cannot be addressed within the confines of short-term thinking. They require a shift toward **deep time** — a mindset that stretches beyond decades into centuries, even millennia.

Designing for deep time is not science fiction. It is a moral imperative.

What Is Deep Time?

The term "deep time" originated in geology, describing the vast temporal scales of Earth's history. But in design and policy, it refers to the practice of planning for futures that extend far beyond the present — considering the needs of people, ecosystems, and cultures that may not yet exist.

Deep time design asks:

- What will this building mean in 500 years?
- How will this policy affect unborn generations?
- Can this system adapt to unknown futures?

It is not about prediction. It is about **preparation**.

Tools for Long-Term Thinking

Designing for deep time requires new tools and frameworks:

- **Scenario Planning**: Exploring multiple plausible futures, not just one forecast

- **Back casting**: Starting with a desired future and working backward to identify steps
- **Legacy Metrics**: Measuring impact across generations, not just fiscal quarters
- **Temporal Mapping**: Visualizing how decisions ripple through time

These tools help designers move beyond urgency and into **durability**.

Intergenerational Ethics

At the heart of deep time design is a question of justice: **What do we owe the future?**

Intergenerational ethics challenges the assumption that the present has primacy. It argues that future people — though voiceless and invisible — have rights. They deserve clean air, cultural heritage, and functioning ecosystems.

This ethic is embedded in the **Future Generations Act** in Wales, the **Green Legacy Initiative** in Ethiopia, and the **Vision 2071** strategy in the UAE. It is also reflected in indigenous traditions that honour ancestors and descendants as part of a continuous moral community.

Designing for deep time means designing with **care**.

Embedding Time into Architecture and Policy

Some institutions are beginning to embed deep time into their structures:

- The **Svalbard Global Seed Vault** in Norway is built to last 1,000 years, preserving biodiversity against catastrophe.
- The **Long Now Foundation** is constructing a 10,000-year clock to foster long-term thinking.
- Japan's **Ise Grand Shrine** is rebuilt every 20 years, preserving craft and continuity across centuries.

These examples show that time can be a design material — like wood, steel, or code. It can be shaped, honoured, and sustained.

The Role of Art and Storytelling

Deep time is hard to grasp. Its scales are vast, its subjects abstract. Art and storytelling help bridge this gap. They make the future **feelable**.

- Indigenous oral traditions pass wisdom across generations
- Climate fiction ("cli-fi") imagines life in future worlds
- Public monuments and rituals anchor memory and aspiration

Designers must collaborate with artists, historians, and philosophers to cultivate **future consciousness**.

Lessons for the World

Designing for deep time is not a luxury — it is a necessity. It teaches us that:

- Progress must be measured in centuries, not just speed
- Justice must include those who cannot yet speak
- Beauty must endure beyond trends

In a world obsessed with immediacy, deep time invites us to slow down, look ahead, and build legacies that last.

Conclusion: A New Horizon of Responsibility

We began this journey by confronting the tyranny of the present — the dominance of short-term thinking in our politics, economies, and technologies. We've travelled across five regions, each offering a unique lens on how long-term thinking can be embedded into policy, infrastructure, and ecological planning. And we've explored the philosophical and practical dimensions of designing for deep time.

Now, as we stand at the edge of this horizon, one truth becomes clear: **rethinking progress is not optional — it is existential**.

Progress Redefined

For centuries, progress has been measured by speed, scale, and profit. But these metrics are no longer sufficient. They fail to account for the slow violence of climate change, the erosion of cultural memory, and the silencing of future generations.

We must redefine progress to include:

- **Resilience**: Can our systems withstand shocks and adapt to change?
- **Equity**: Are we designing for all, not just the privileged few?
- **Legacy**: Will our actions be remembered with pride or regret?

Progress must be measured not only by what we build — but by what we sustain.

Cultivating Future Consciousness

Future consciousness is the ability to think, feel, and act with the future in mind. It is a skill, a mindset, and a moral orientation. Cultivating it requires:

- **Education**: Teaching history, foresight, and ethics as interconnected disciplines
- **Institutions**: Embedding long-term thinking into governance, finance, and law
- **Culture**: Celebrating stories, rituals, and art that honor time and legacy

This consciousness is not innate. It must be nurtured — through dialogue, design, and deliberate practice.

The Role of Leadership

Leadership in the age of deep time is not about charisma or control. It is about **stewardship**. It requires:

- **Vision**: Articulating futures that inspire and unite
- **Humility**: Listening to those most affected, including the unborn
- **Courage**: Making decisions that may not pay off within a single term or tenure

From Wales to the UAE, from Finland to Ethiopia, we've seen how visionary leadership can shift the horizon — from immediate gain to enduring good.

Designing for Inheritance

Everything we design — from cities to algorithms — is a form of inheritance. It will be received, interpreted, and lived by people we may never meet. This reality demands a new ethic:

- **Design for repair**: Can our systems heal what was broken?
- **Design for continuity**: Can our creations be maintained and adapted?
- **Design for reverence**: Do our choices honor the dignity of life?

Inheritance is not passive. It is shaped by intention.

A Global Movement

Across the world, a quiet revolution is underway. Governments are appointing future generations commissioners. Cities are adopting 100-year plans. Communities are restoring forests, reviving languages, and reimagining education.

This movement is not uniform. It is plural, contextual, and evolving. But it shares a common thread: the belief that the future deserves our best thinking, our deepest care, and our boldest imagination.

A Call to Action

Let this book be more than a reflection. Let it be a **call to action**:

- To designers: Build with time as your canvas.
- To policymakers: Legislate with legacy in mind.
- To educators: Teach futures as a form of citizenship.

- To citizens: Demand systems that honor your children's children.

The horizon is not a limit — it is a responsibility.

Final Reflections

We are the ancestors of tomorrow. What we choose today will echo across centuries. In our hands lies the power to shape futures that are just, beautiful, and enduring.

Let us design with courage. Let us govern with conscience. Let us live with compassion.

Beyond the horizon lies a world waiting to be imagined — and inherited.